Nobody would believe those, if they actually told anyone, what they witnessed my getting up to with their own eyes.

With no sleep; nor water: nor food for a week, during martial art practice, I came upon, a discovery, possibly only due to not-blinking, while dry-fasting.   I came to see with my own eyes, what women, actually see, and how men and women's interconnected problems stem from sight itself and occur cyclically starting from childhood on. To overwhelmingly empirically evident are the truth's and their synergistic lies, for anyone to refute. And if anyone does refute, I will be happy to clear one's distortions up publically for them. Without wasting anymore time.

Let's see why humans, women and men all fight, always, without any exceptions,.

I0480457

- Conflict bias
- How dare women accuse men
- No double standards I know your cheating
- Biologically biases are grounds for conflict
- How to uncover another bias
- How to be aware of what's needed to re-evaluate a bias
- Monitor real internal yet physical manifest effects of bias
- Mum you didn't look I saw you not look why didn't you look

- Mum over here
- This isn't mars and Venus.... This is just peripheral and linear, vision, and behaviour, and concepts, and goals, formed either peripherally or linearly.
- How much does the sense of sight effect all of ones behaviour

Conflict Bias?

IntroductionDis-cerne-

Discernment is a word from the late fifteen hundreds.

Discernment = means the keenness of the intellectual perception; for insights,: acuteness of judgement.

"Dis" = "to detach" to take "apart from".

"Cern" = from Latin, cernere, meaning to sieve.

meaning to separate; set apart; sifting through: intellect.

"Ment" = from the late sixteen hundreds means the act of perceiving via the keenness of intellect.

Evaluation is Latin, originally from mathematics.   Evaluation

Evaluation, means, basically, to making calculations, which are of the best value, from the word évaluer. 'É' = "out" and 'valere' means "value", take out the proper value from any problems', no matter how many people and problems are part of the equation.

Evaluation basically means: "the best way to better calculate one's benefits and calculate values of", literally meaning to "obtain the value out of", "take out the value (of)".

Evaluation equates to discernment plus appropriation, and is the immediate result of evaluating, via discernments', thus automatically appropriation, occurs, as discernment itself and appropriation itself are outside of the realms of a bias.   Appropriation .

Appropriation means

To add up something to make one's own property.

To conform with English grammar laws to add {add = app) (something) to make your own" from the word, proprius from Latin, meaning to make one's own. Were we get, the word proper, meaning to adapt to fit (some) purpose, aptly with commendable excellence.

Culminate from climax [astronomy], means to reach to height (of something).

Bias, is a word, taken from France, five hundred years ago.

Bias

Bias means to "slant/slope" or "to incline to one side" (and thus rendering the attainment of the others perspective, from the hill's other side. By both coming to the middle, both become able to see both of each others sides. Other wise, any biases, prevent over standing of the adjacent sides perspective is rendered unattainable.

peak, of the slanted; slopes: precipice, to move

the other side is obsolete).

Conflirge

Conflict = contend; struggle; fight and wrestle.

What do you think about women & men seeing two totally different version of reality, to make harmonisation, based in just one agreeable reality.

When women and men, clearly see totally different versions of reality to one another, conflict bases are set, and each sides point of view is imperceptible, until all conflict and mostly all bias, is removed.

You can see why my next book has already sold copies , doing what no one else can, creates a niech situation.

As no one else you know is talking about this, because no one knows, to ever go as deep, as my five books series does. Kings with the wisest advisors, haven't known this. thus you can't get this anywhere else, right.

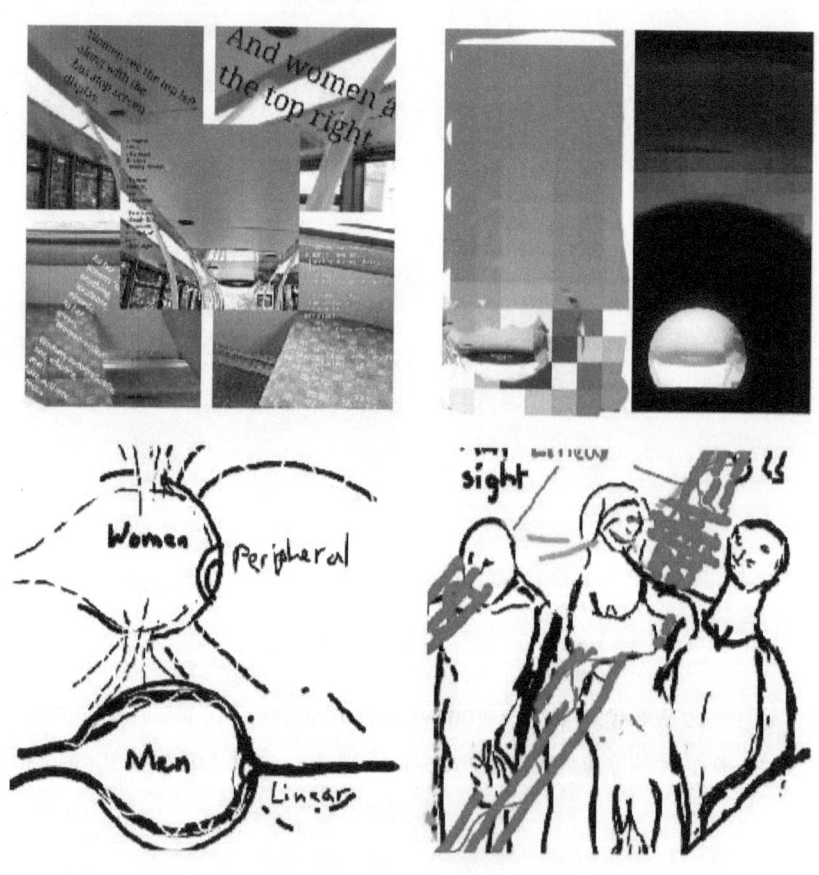

How dare women accuse men of what all women do to all men women accuse..

Top left, look at the way a women automatically, sees all angels, when also looking at what the bus display says

Top right

Top right - is how looking directly at the bus display screen, at the same instance as the woman, looking at the same displayed screen, (every woman, all women see like this, not just some, and instead all).

Men's sight – globally is a telescopic, kind of sight.

Bottom left women

Bottom left is how women and women see, From women's eyeballs, all angles light can make it to women's lenses, make the eye automatically process, optical stimuli, giving panoramic vision, all around, virtually, and with all immaculate visual definition, ultra HD.

This is why women never turn to look at men directly, because the direct look is the dead giveaway, this allows all women to thoroughly ogle you men as much as they like.

This is also why women don't over being directly faced, they externalise men have their vision, this is why women accuse men of cheating whenever women and men go out somewhere together.

look at the left women pictures and it's, obvious why they will, externalise this about all men.

A woman's idea of staring is a man's way of seeing.

Cause of little glances , were a man, [bottom left men], attempts to form a focused image, to even start absorbing women's image, men are immediately, in tern presumed to be sending strong sexual signals, by every woman world wide, even family members do this, yes all non immediate family members (all cousins) actually believe that a glance is a sexual indication of storing undeniable magnitude.

Obviously. Look at the left images, for women, they see all around, perfectly clearly, if men saw so clearly, all around, men would not ever, need to turn to face a women, to make her uncomfortable.

Yet only women get to ogle thus men must turn.

The issue is, women, if any woman, globally, glances directly at a man for a second or zero four seconds, this is a strong sexual signal, that women are attracted and want sex with said glanced at man, women may glance at fifteen men in a minute or less.

Now with men's sight, a glance, doesn't even allow our image, to focus, sight not brought to focus, is not visibly decipherable.

Thus women globally, need to also be made aware, of the fact, men's glance is no way sexual signalling, this could be for several reasons, yet believing have already seen women, without directly looking is psychosis.

Now tell me something, if you put a telescope to your eyes, and look, around, upon, seeing a person, do you not have to move the lens of this telescope, around their body and tightly connect the images, you see separate from one another, to build and grow a concept of the observed.

Of course, you need to tightly connect the dots, right, of course.

If women, become aware of this, men which work decades, to build homes, house the family, drive the family, feed and cloth and support the family, by working three days per week, with no sleep at all, working through out the week for decades, for years and years.

To come home and be accused, by a women's projections, that men have women's vision basically, when men do not.

Means the women are cheating, mentally and physically, at least on the biochemical level, at the very least.

Women are all cheating, and so much so, women abuse men, because women are so prone infidelities.

Later we will cover the pattern which women come to fall in love via, and let me say now, if a man does not take her virginity, she does not love him, she is still in love with the virginity taker, this is another reason, women will abuse the emotional crap out of a man, and not care, men have scientifically, verified, heart tachyarrhythmia (irregular heart beats, because of the strain women put on men).

Now if men, are allowed to see societal populations, with out being narcissistically attacked on a double standard basis, than men have but one way to do looking with.

And men don't have women's sight.

you can see the picture for an immediately absorbable representation of what women (all) see. Women believe all men see this also, thus you can see how women's behaviour is shaped, from this visual diagram shown.

This is why women, from a stand point of bias, believe men are being perverted, and coming on to strongly, just because women, see all men's penises, bulge through their jeans outline, so women don't require no head turning to see a whole bunch of crotches, all at once.

Thus all women, get very offended, when a after absorbing a thousand crotches a day, at default, one man decides to turn to see a woman, to absorb her image and demeanour, just once.

Women look at a thousand or more crotches each time they make a commute from one side of the city to the other, as exemplified via the top left image, can you imagine being on the train line, from start to final destination, during rush hour, millions more crotches do women absorb then men will, per life time, by default, this is a hundred percent consistent across the world.

Men may not look at even one woman, some men may look at a few women, due to innate chemical attraction, curiosity about a woman's demeanour, can make a man's curiosity, choose an innocent reason to look, just to see, not to talk, nor signal nor fantasise on any sex.

A women, will interpret this innocent action, as a full blown, inappropriately, hard-core sexual signal, and all other women, in any public, space (globally) will concur, that said woman's, said man, was a male slut, for looking at another woman, for 4 or several (7) seconds.

These women are not aware, yet once made aware, does everyone recognise, women will become narcissism, by default, if women which see every crotch do not allow a man to look at one, knowing now, why men must turn to look to in tern see.

Women = thousand fold more crotch watching.

Men = thousand fold less crotch watching.

(Men versus women: one thousand fold the trouble, for looking a thousandth less).

Even while under men's protection she's cheating, ogling, conceiving cheating and cheating, and you can not catch her unless you know this or unless they know this.

Women accusing men intently of cheating, are cheating themselves, and the phone probably has the information right now, take it get it unlocked, and get out, don't give away half your life's work, half the children and everything else, get out, early.

Don't forget, each adversity has an equally or greater seed of benefit, the next book will show one how to choose excellent partners, especially online and in person.

Some men don't even look at women's crotches, some won't even look at my other women, for months or year's, until being accused for so many years, it seems easier to stop self abuse, were women use men's delicate feelings, as ammunition, to project infidelity women exhibit, on to men.

No double standards.. I know your cheating...

Men eventually, do what women tell men, they've already been doing.

If men accuse women three hundred times a day of cheating, women will, or of being ugly, women will feel ugly, someone accusing men of cheating, says, to men, I believe your cheating, always.

When which accuse men, this means by dual rights, women presume to be free to now cheat, just by the act of accusing, ne-gat-ivism, right.

When women accuse men abnormally more of cheating, that's when women actively are cheating, women than project their guilt on to men, because when men feel those emotions, women need not feel those emotions them selves.

Women always actually never stop seeing all dick around.

Women with sunglasses are the worst.

women with sunglasses are narcissistic, nine times out of ten.

Their eyes don't need sunglasses already to see, acutely, yet sunglasses, allows closet lesbians to ogle other women, as women are highly perceptive, thus women use sunglasses when scouting other women; are closet lesbians, and narcisstic.

Women with sunglasses often drink loads, thus the wisp of delicate male emotions, aware completely lost from female vision, from just one to two beers, by three women are ready for bed, with a man or without, yet get out in the mourning, drinking means women, will kill men emotionally, just two beers a day, is highly dangerous, narcisstic women, (make seventy percent of al narcisstic people on the entire planet, yes two thirds of narcissistically developed people are always women, and two third of these narcissistically developed women, are always said to have two to three beers per day, new facts, as researchers discovered, women were using ne-gat-ivism, amazing right.

Women are so narcissistically developed, they conceal, their inhibitions, so collectively, males were thought to be the narcissistic one's, contrary to the newly proven truth.

Alot of women which were talking about narcs on YouTube, are themselves narcs, and have some kind of goitre like frog in their throats, every time these women attempt to say anything about male narcisstic personality disorders, because women know this is actually them, me these women, use ne-gat-ivism, yet the internal conflict by talking about women's own bias, as narcissistically disordered people, seems to create a goitre like frog, the size of Kermit, in their throats, and I know why, I can tell what they read.

Men you are not narcissistic, if the Greek Marc, saw his reflection and love himself, in water, he was likely an empath.

Women see twenty fold more ,and have twenty fold more make up products and cosmetics than men, women have one hundred more shoes than men.

Women lie way more and operantly condition men, from men's protection, women ogle other crotches, and condition men not to be scared to look at another woman, women take men's money to financially abuse them, men don't take women's money,

men don't look at other women, and than impose double standards.

Men don't take women's children and smear them to the world around, it's the other way.

And women ogle every man and women in ever public space without fail.

By inception a very unfair double standard.

.

Bottom left, demonstrates, why men have no concept of cheating and don't know why women accuse men, because men have linear vision, if men were directly facing and pointing their eye at them, men were not looking,

Just like the earth sees the sun, while being mostly hidden, yet the sun (men) have to do serious focusing, to see earth's beautiful (sur-)face.

When men attempt to see a women, she gets very uncomfortable, with the angles they use, yet, you know women look at every all men's crotches, they pass, ever, virtually.

Vindicating men which themselves, wish to look for a small while.

Men are not being rude, she 's only uncomfortable, because women don't know, in fact how men see at all.

women presume men directly look to say I want sex with you, every time men look, thus women think men just go around all day, looking at every women to say let's have sex, because no women, know this, none at all.

This is why every mum thinks her son's a narcissist, when he says, "mum, directly look at me head on, while I do new tricks, by looking at the top right picture, you can see why no boy saying this is ever being narcissist, men all externalise this sight on women, globally.

This means women sight makes all mums externalise their own sons are narcisstic, because in fact, men have to what women do to, it's just we can't hide it very well, because men do see as well as women see, clearly.

is sexual by directly facing and looking at women, even in the very slightest of glances, the top left diagrammed images show why women mistake each little glance as a big sexual sign, you see now why when women do decide to look at men, this (globally) means women want sex.

Show them this image and explain, why you have to look as men, they will chill out when they see the picture or hear your explanation. Remember, women are externalising and cross calibrate their own sight, on to all men globally, yet this, still women's sight, not men's sight.

Cinema – when your girl tells you to not look at the actresses on screen, tell your girl:

if you just saw me from facing forewords,

While eating snacking on popcorn and drinks,

While in the dark,

while watching the movie,

Simultaneously connected me looking, and saw also the women or women I looked at,

All at one instance, with definiteness in your voice

than you looked at every single crotch we ever passed together

and you need to stop seeing any other members of the opposite sex now, if you want limerence, or to not imposed, what by inception a double standard.

Send her to my website, have a telescope, when she goes to the movies, instead of running your fun, while she gets her perve on, she can see the movie the man sees, or she could get the balaclava hood.

Telescopes, get women's own eyes (which women take as proof of truth) to actually teach women, women are the only promiscuous, one's innately of men and women, and this is limerence.

Bottom right

,The last diagram, shows how men have to look at women, and the small amount of detail men can actually focus on and see world wide, all men see like this, thus men stare loads more, because, they see way less.

For women, you must wear a balaclava hood to a bar or party to actually see, how much more, than any man on earth, women, how much more than all men on earth, women eat crotch with their eyes, like chlorophyll eats sunlight, it's actively actually very intrusive, when you are the only ma, which perceive women's cult like behaviours, and rituals of negatively negating, all the dicks you strive to look at, while calling men dogs, projecting women vision on men, what's more, women all compete, to attract the glance, which women, upon success, have only validated themselves as expressing whoredom.

Actually no moral high ground nor moral high horse, for any women anywhere, world wide, to ever call any mean male sluts or dogs, actually exists.

Of course in a women's reality, based on being educated and indoctrinated via a women's own sense of sight/vision, women have felt justified, to accuse men of being dogs, of course women have, just as boys, have felt justified to force mummy to look directly at them.

Yet in actuality, the fact is. Cross calibration, has never, allowed a balanced, fair, ground, for either sex, to evaluate a love strategy.

Based off of actuality, as opposed, to contrary and conflicting and juxtaposed opposite, sense for sight and vision respectively to men and also respectively to women.

You must study the GOJUVISION YouTube channel, to discover, the highlighted points, at which women and men are essentially non-verbally communicating and the externalisations being made.

This has never been done, coming over to my channel men, means you will learn the secret language, women speak with and see with, and men will understand, and men I'll understand all those things which makes them uncomfortable around women at once.

Women will learn all the aspects which make them unappealing to men.

So women unfortunately, cannot afford to miss this, at all, as women are appalling at keeping men happy.

Contrary to what you'd like to believe the evidence is overwhelmingly indicative.

Because women have mis-externalised, that's men's glance is a sex signal, such women, please buy the telescope, and the balaclava hood, let your sense of vision, teach you to see, as men do.

Because one's own eyes, will be believed as proof, women are accusing men,

Staring fifty fold more than women, result in men, seeing fifty fold less than women, at least, yet imagine if men do no looking at all, this is societies atrocity, men you can look your not a pervert, you could never be, comparably, they see to much for you to ever catch up.

And women discuss you to women they do not even know, all the time, as soon as don't look, like those Luigi's mansion ghost's, that's what women do, once you look.

Biases lock out evaluation

When Chelsea and arsenal are laying off at the games, there alot, ongoing on, biologically, biases are grounds for conflict.

Biases, do not want to move, thus biases do not evaluate, something in their best interest, even if self evident

Your discernments' are innately self-educating; once you adapt; anything you've sieved; through, via your keenness of intellect; to best fit some purpose/s.

Repeating, the same discernment and appropriation five times (consecutively); contextually conditions, one's long-term memory pattern.

The shaping of that thought (memory) pattern, in tern, shapes the locations and trajectories, which discrimination, can occur from.

Main

The neuron itself uses norepinephrine, to make the neuron itself work, the neuron in tern is conditioned by neural-use.

The neurons make the entire brain, the neurons controls one's entire body, so one's neuron in tern programs (conditions) one's being, like a puppy or programmed can be conditioned or programmed.

When used, the neurons, thus relay, the bodies reactions and the brains reactions, and make those the neurons new conditioning program.

Norepinephrine and dopamine, help lock the left hemisphere of the brain, to keep doing any practice, keep carrying any brain baggage (thought/belief/emotional pattern), anything the brain and body has already practiced, thus been already doing, is were one's standing (albeit bias) lies.

Norepinephrine fixes and locks the brains' left hemispheres', into a culminated bias, (as a physical effect in the neurons chemicals*, and) as the neural pathways keeps one alive, self preservation, one's, first natural law, thus basically survival.

One's bias thus just seeks to discriminate; in accordance with one's bias(e's); which primally and primarily are just a basal urge to survive, against, the bodies perception for an attack, however arbitrary the perception of an actual attack(s) may be.

Basically, the human brain and body, condition's hardwired schematics, with your neurochemicals, as the code that programmers ones' minds' (and bodies) conditioning's'.

Unintentionally or otherwise, opposing one's biases (such as supports/beliefs/groups) results in thus hindering the practiced appropriations of neurons working people's bodies, the same interconnected neurons responsible for calm homeostasis, digestion, breathing, releasing adrenaline and so, is akin to disrupting one's entire biochemical equations' equilibrium, as much as if one were directly hindering another's neurons practice of breathing, walking or blinking.

And indeed, the fight and flight response, is only triggered via disturbing one's biases. If every thing one desires they come to have, one doesn't get angry. Having what one doesn't want happen can produce the opposite state of being. One supports one's own biases this is called self preservation and survival and is the first law of nature.

The brain and body, perceive arrestment (hindrance) of a culminations' attenuations', as a threat (or virus) in their neural-circuits. Matrix of One's beliefs, equates to ones survival, and thus, in defending the side, of ones perspective, one may attempt to thwart you, to re-establish relative peace within.

Disruption to one's neural cycles, practiced attenuations, disturbs the cyclic pathways cycle from being complete optimally, creating, elevated heart and breathing rates perspiration go up digestion shuts off stress hormones elevate and the overall internal destabilisation can lead to conflict externally, to seek a survival orientated remedy from one '(s)own standing from which those actions are an auto-attack due to one's standing.

An arbitrary attack, is an autotelic attack, the conflirge makes one feel less and less good, thus minute biases may be focalised and pronounced, autotelically, things that feel good are self perpetuating.

An automatic attack, is due to standing, one's ground's, one's side, form one's bias.

Thus look at a football game, (midrange attackers and defenders) once a side is picked, the attenuation is automatically autotelic, if all goes to plan, everyone feels good, if all goes to shambles, everyone feel autotelically attacked, as one's automatic attenuations 'are inhibited, this obviously extends to one's beliefs; which teams one supports; comforts; and ultimately one's very own practiced common standard.

Anywhere one perceived conflict, one also perceived, yet fails to notice, the true side and standing, of anyone at all, is revealed. Conversations which are open, allow one to access their favoured topic's, empowering them in and supporting them with that which is already practiced and favoured (said topic), so, creates, mutual time, spent on mutual grounds, thus your virtually already dating sensually, as internally, the hormonal states, connecting the two of you, are established.

At a date: Another's Favourite topic is not always one'(s)own favourite topic.

Any topics, which she's gung-ho about, are most cathartic for her, thus are your favoured topics, as they make her comfort occur specifically to you, when her topics are shared with you, your topics don't have to be discussed, at all.

Do this and you win every time, even when you lose, it's only because you weren't motivated enough to do this, through and through...

Just talking to girls is the way to create oxytocin and make them feel safe to be vulnerable with you, fifteen minutes is good, breaking touch barriers can be good, just talking for fifteen minutes is the goal, drop the bar, don't pick it up, go to zero sense of social normality, and speak as you have words occur, don't dilute, nor alter the word's consistently, nor focalise, drop the bar until the bar is so dropped, that no actual way exists, thus anything said or done, occurs via magnetism and not force, also do all this, with perfect erect yoga posture, use females characteristically receptive gaze, face aspects of your environment to look considerate also go to Go-ju-Ryu, to learn and practice better peripheral visual capabilities, which will keep you calm, as peripheral vision, occupies the vision, so dilutes and de-focalises intensity and increases awareness, so propelled anxiety to plummet.   By default drunk women, are automatically aroused women, psychologically, the same not so for men necessarily.

Be present silently, yet, physically present, warmth; breathing; aura: as well as talking, both attenuate practice to their neuron and yours, so your actually "in there", in the brain, the mind, and the heart, if it was that kind of meeting, even the pallet.

Arousal of one's internal conflict which gives rise in tern to one's external conflirge for seeking remedial peace, imbued's oneself with empowerment, as the remedy, is the antithesis, of the presence of a bias conflict to one's own.

Meaning a state of love, is the default empowering state which empowers all and disables all others, conflictive behaviours, love empowers people to be exactly who and what they already are. As seen with babies taking steps, empowering anyone, from exactly where and what they are already, can have potentially, limitless results.

Summary

Physically, biochemically and neurologically, supporting biases, (which includes not being a threat),allows one to evaluate you as one to form an empowering bond with. Go with another's practiced attenuations and their quotients of emotion and anyone will take you in as one of their own.

Merely avoiding touching on one's biases, will in of itself, avert the ground on which conflict can even occur.

Visually and conceptually, you become recognised as one in the same as another, via their biases. Which are very deep niche foundations, difficult to orchestrate otherwise.

Thus amongst any group of peoples, anywhere on the earth, which can become inclined towards you, via simply, appropriating around one's biases, which empowers one to one's innate inclinations.

To have the
most impact in one's goal, set out for them, to achieve one's
attenuations', one must circumvent, having one's efforts usurped via the
disruption conflict incurs.

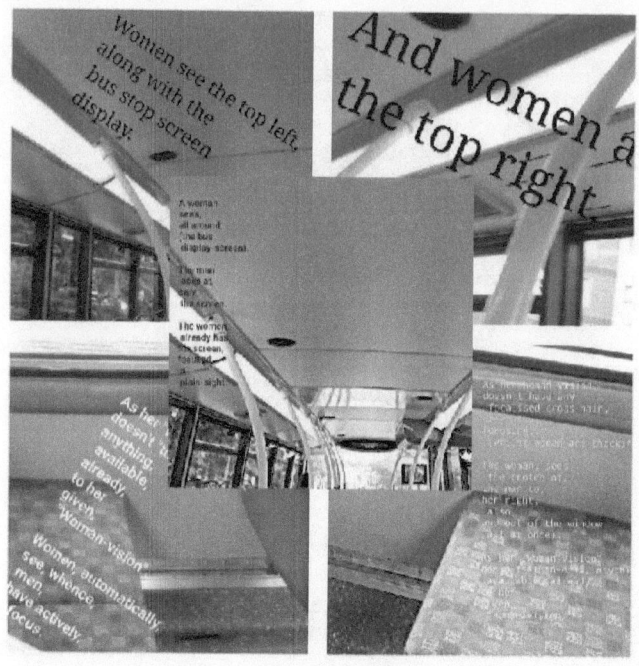

{The right is visible to all men}.

The {above left} is visible visual all women have.   Men see, and so take their sight to be

evidence of the definition of sight, thus

men as sons, have been very forceful,

to have women focus closely on them,

which comes across as inconsiderate to

women, men presume to focus women

due to men believing women see as men

{See above right}

26

Facing the front, at the bus display screen,

Women are granted vision for just over 180°,

Women take the vision they see with as the,

Experience which defines what visible sight\Vision

itself is, women are taught sight by their own,

Optical neuron, which obviously, is sensory education about,

What sight is itself, thus women have had to believe,

Everyone sees everything, which they do.

The left image, is why women get pissed of with men, when women me men are out, women have to much to society around to appease, and live up to, and get validation from, which is why women get pissed off, when outside, and men get pissed of with women at home.

Living for external validation is the basis of narcisistic personality disorder, virtually all women by default do this, way more than man, because women are visually, way more aware than men.

Women are narcisistic, men are not.

Equally, women are very seldom ever empathic, around two decimal percent of women, even possess truly empathic traits and inclinations.

The above left image demonstrates, why women, never see empathic nor non empathic scenery as magically intensely and focally, more accurately, as men, innately do, thus more male empaths exists, and female narcissistic women, pretending to be empathic.

Go to GOJUVISION – YouTube; were I break down empaths and narcs, for men and women respectively: in two minutes.

Conflict biases, different sides to support, breeds grounds for conflict

Intro

Conflict is principally, a synergistic component, attenuated to bring about peace.

Great wars are said to bring great peace after the conflirge, as future conflirge, uniforms attenuative conformity, making the spoils, more pronounced.

Somalian and Indian and black nationals which fought in both world wars exemplify this as citizens of a victorious nation, the spoils are more abundantly available.

They were in the second war and many of their descendants live in ethnically mixed areas of the victoriously developed (first world) nations.

Main

Weather passively; unconsciously: or otherwise, directly, one will attempt to thwart anyone opposing their attenuations' and biases, without any examinations of the quality of value of one's biases verses one's another's opponent's opposing biases.

Whether ones attenuations serve them any longer, quickly turns irrelevant, as evaluation, does not fruit from biases and conflict does fruit out of biases.

Discrimination fruits from biases.

Some overcome biases with conflict and others overcome biases via evaluations, which results in tentative peace and harmony, with out the presence of conflirge.

While on the other hand, Discernments', alleviate discriminatory tendencies, as the qualitative values of one's biases, are evaluated for their usefulness, by the very act of discerning, discernments.

Discrimination, seeks to take things apart, to reject and disharmonize, in accordance with relieving (balancing) ones inner conflict.

Discrimination, seeks to reject, and even can go to the extent, of including physical violence, to alleviate internal conflict felt at the result, of taking apart, something in a different and ultimately conflicting nature to one's own accordance, with one's own nature.

Biases arouse contention and thus are easily identifiable and biases clearly lack evaluation, thus may be poorly rationalised, upon evaluation, as biases are reactionary reactions, so no space nor time, for any form of evaluation takes place.

This conflict biased-effect, is evident in, Chelsea versus Arsenal, style scenarios, a reactionary state, in accordance, with one's own sides goals.

When another's beliefs and views, are not in support of one's own team; (or not in support of one's (s)own beliefs)), conflict is the natural insured occurrence and the normal outcome, at any instance's' conflict, (which means to contend; to compete; or to struggle).

Externally bias aroused conflict, may be direct or in direct, or even unconscious.

Summary

You discern to learn, to appropriate, what you've learnt, via your discernments.
Repeating, discerned-appropriations, over and over again, forms a culmination. Culminations favour anything in support of their conflict biases (thus disregards all else).

Thus Anything that poses friction, to cultivated culminations (which are, the height of your discernments' appropriated biases), thus arouses a (second nature) bias.

Thus to unbiased any situation is to remove an incline, to just one side. Thus biases disregarded other information, not on the same side of the bias/es.

Only serotonin mediations, enable the evaluation and expansionary appropriations, attenuated by the support of love, which de-culminates, otherwise the inclinations automatically generate contentious responses.

Conflict over ones inclinations, is indeed a conflict bias, as inclining to one, (one's) side, occurs via slope's.
The word conflict, came from Latin over half a millennia ago.
Conflict = contend; struggle; fight and wrestle.

Thus a conflict bias, is whence any contending struggles, over respective inclination towards contending sides occur.

How to uncover another's conflict bias

Intro

This conflict is actually due to pre-existing biases, resisting contention, via persisting into their bias and an externalised strife called a conflict bias, arises.
Opposing conflict biases, causes contention (strife), thus you can see what someone's attenuated to.

Main

33

When contention arises you've directly touched on a conflict bias or around a conflict bias, this to tell you exactly were or approximately were, a conflict bias lies.

Biases lie around, exactly what one themselves is already, preoccupied with doing. Doing something makes the neuron, derived, the bias, doing is basically developing, the mind, the body, one's entire self, via just taking steps, walking and eventually running become possible, from just those steps, doing developed bias.

Not doing de-culminates bias.

Obtain one's conflict bias

Walking into a group at an away game and saying, I love Arsenal, reveals any Chelsea supporters, even if not vocalised; to be in support of Chelsea, one's same neurochemicals operate the entirety of one's body; also control respiration and heart rate; twitches and triggers (due to the strife); occurs as physiologically, manifest internal conflict, of being on the bias of supporting one s side.

Monitor real internal yet physical manifest effects of bias

Ones physiologic, instinct for self preservation (which is the first law of nature) is directly disrupted and effected, via conflict biases.

Internalising conflirge activates triggers, through out the same neurons cyclic

functions disrupted (disempowered) with the dis-crimi-nation.

Thus this neuron, primes and conditions all of the proximal, neurons, and thus the entire body. .

Emotionally detaching with one's own, auto consciously welcomed attenuations', which are welcomed distractions; from what's no longer an secure empowering location, offering serene environmental encounters, indicated via escalating attenuations 'to conflict fashioned around separately supported teams and aroused via a separation of biases, leading to conflirge.

Outro

After you have a conflict bias, learn those fight and flight triggers, which reverse, engineer the triggers in others and tell you were a another's biases lies.   With out you having any knowledge or awareness of such bias, outside their gift and flight response, as days go by.

Suddenly getting up; (is in fact, a somewhat nullified, fight and flight response, yet a fight or flight response none the less); moving away, indicates (flight') meaning another doesn't want to have any contentious issue/s.

The environment their in; is likely the factor; work; social: family, may, in tern limit expression of one another's, clearly, nullified reactionary, flight and fight instincts.

Intro

Appropriation around evaluated, self evidencing facts, is not the same as culminating around a conflict bias. Discrimination is bias (as inclining in favour of one side, is bias and discriminatory) and discernment is combining both sides to best place the evaluations on ongoing basis. One (dis + cerne + ment) suffices in being evaluation, which ones discriminations, utterly lacks.

Main

You take food, to a dog, they accept the food with you, this is a discernment, they can clearly sense, your attenuating support, your presence is thus supportive.

Take the food away from a dog, which you did not present, discriminately, arouses conflict, you are culminated around the displacement of their eating of their meal.

For an mammal or human, to take the value out of your interface-presence, calm is required, otherwise, this attenuation, garners lasting dis-crimi-nation.

All men\women\mammals, come to be via empowerment, as all are vulnerable from birth.

Outro

Thus conflict, around and over bias, innately, is a enemy of the critical staple, thus men women and mammals, discriminate against dis-crimi-nation, via dedicating the neurons to survival, for example discriminating against predators, makes one a better member of the heard, dis-crimi-nation against potential threats, and discernments', makes one by default, valuable.

As a culmination, is difficult to automatically de-culminate or re-evaluate. As a conflict bias, is automatically attenuated for only cultivating support. Biases otherwise, discrediting or discarding, do not take any benefit, anti-attenuated biases.

How to re-evaluate a conflict biased

Intro

Unconditional Love is support, not attenuated to a conflict biased, love which is unconditioned, thus conflictive, is thus supporting, yet not geared to nor from a conflict bias.

Conditional love based on conflict biases, can expand, to unconditional love, if sufficiently evaluated, via culminations and attenuations. Unconditional love is, and conditional love can both lend themselves to ambivalent strategies, for transmitting that respective love with, respectively.

Just the free flowing, force which expands awareness.

Love itself enables learning. The catharsis of love provides the attenuations' framework, which is a state of being , sufficiently comfortable, to prompt exploration from.

For some people, unconditional love can be violently beating your partner, as without any conditions, this can be with in the permissible limits of unconditioned love.

Main

If you have not advanced in a topic at school, the problems normally lies in a lack of good-diet; exercise or most commonly and most probably, the love, to attenuate the culminations of ineffectiveness, to be evaluated, into the right frontal hemisphere, via evaluation you can take apart and calculate the value, of the previously un-graspable (un-learn-able) topic, and best fit any purpose, which ones unable to learn or grasp.

Teachers don't normally love their students, and vice versa, and thus this kind of beneficial catharsis, which greatly accelerates learning in adults and children, is often lacking.

If you think about arsenal versus Chelsea, the conflict occurs from different sides, contending, whilst, love accesses, zilch biases, no sides exist, the supportive foundation, means anything you do, via inclination and autotelic attenuation, means one is always supported, to do anything in any state or condition.

This means, one does not have to worry about triggering biases which leads to conflict.

Rather ones, own innate explorative compass is free to attenuate and point one in the direction the desires has to have, this is why, children and adults, do fifteen percent worse when doing exams in foreign rooms and buildings and even more so why another thirty percent decline may occur, when a different tutor or teacher is used as an examiner.

Attenuations, of children, especially boys, are the most bold and brave and free and jubilant, in the presence of their beloved mums, and less so in the presence of all others, as their mum, supports them the most, out of unconditional love.

Outro

This means, in summary

All mammals, all men and women, depend on empowerment, as a means for survival itself, and thus this harmonised neural code is an impetus to a consistently sought goal across all surviving culture's and creature's and species, and has been, since the earliest, men and women, originally and initially, existed all together.

Empowerment, from the womb, all the way up to, walking independent women or men, is a lineage of empowerment for nurturing one's developmental growth. Sometimes empowerments sought via conflict, and is the human natural (basal root chakras) survival attenuation, and in tern, is humans surviving legacy.

Love is the most empowering of all forces, the most powerful of all forces, is loves empowerment, love.

Love liberates the conflirge, which results from the "biased-effect", by activating neurochemicals, which decode (and overwrite) and avert the programs for conflict, becoming active, as a result of biased-effects. Favouring exception of one's aversion, to one's biases, which otherwise results in conflict, is not an issue in the face of unconditioned love.

peripheral vision

Pink lines are peripheral visions target's (clearly visible)

Women see the world around them, women do not need nor want anything to do with, directly staring, focusing would remove all of women's given visions periphery's range of vision to see all men at once with

So at the bar, she's checking out everybody, throughly, on ongoing basis

Look how many crotches she is spying on virtually all at once

Women take all and every opportunity to look at every man's crotches and buttocks.

Women have seven fold more sex than men do, they see all and thus women want all the more.

Women don't want to let anyone in, they do, yet women need to be activated to feel energised to start their feeling like women won't be trapped with a baby with a un-magnetic, weak, liability (financially), or a man child, and yet women want men to break the ice, and make women "excuses", to be able to rationalise against their fertility minded survival, and the fact women really badly want loads of sex, all the time, the diagram shows why.

Imagine you could look at women sexually all day, and your partner never stopped you, you would be happy and also you would be more oriented due to more of a concept about sex.

Overcoming conflict biases with love

Biases are born of love and nurture, to automatically support such love and nurture, yet biases lead to conflict. Thus biases seek love and nurture, as basically, making sense, to the neurons, which themselves, draw and conceive of these biases.Intro

Romeo and Juliet displayed a trait of love, in which, love itself, will use feeling, to opposes, any opposition of empowerment of one's loved one's, as a default.

Receiving love will thus take away one's bias, reservation and empower one to learn from another. This is a physiologic fact, rooted in activating the right prefrontal cortex, that is used for evaluation.

Main

Love thus, makes learning, from a place of love, results in aversion of conflict biases, and the catharsis activates the (right) hemisphere of the brains for evaluating with, when the human IQ has culminated the highest peak the IQ will come to ever be.

Romeo and Juliet were maleficently, "star-crossed" lovers, in astrology.Romeo and Juliet, exemplify effect that loves presence, has on opposing teams/sides/houses/"zodiac-signs".

Juliet and Romeo went against tradition and culminated biases, with love, the biases, occured themselves, out of respective love, and during conflirge, Romeo and Juliet circumvented the conflirge, which allowed them to evaluate and conceive of attenuating to one another's love.

Like Aires and Libra, most profitable grounds to bring out one's attenuation from is love.

Outro

Love (empowerment), gets spent through, one's innate calling, anything less than love, could have been, empowerment one could of had, for their innate attenuation.

Bearing this knowledge of love in mind, will help you to sieve through this information, better. Toxicity, is born from conflict and principally biases, which are the grounds, conflict grows from.

Why do al boys, need to force their mum to look head on at them, yet not girls, why would mums, just look at boys, with the same vision men has, so the boy wouldn't only have to tell this to his mum.

World wide, regardless of socialisation.

(No one could argue with me, on this, not even Phil valentine) mom's of sons them selves all witness this, and mums never manage to give the exact sight the boy asked for, mums give it serious goes at least once, women could never see as men do, women should have stopped projecting they see as men do.

"Mom you didn't look I saw you not look, why don't you look."

The inner golden compass points and screams out what needs evaluation attenuated.

Toxic cycles of ruminations and murmuring, as culminations, when evaluation should take place.

As murmuring, points out (auto-attenuative grounds), something should be discerned, for enrichment, which is precisely what evaluation('s) for.

46

If the bias inclined to ones best interest, the bias is more accurately, a conflict of interest. Corporations and fiduciaries often lie when their best interests conflict with facts.

Lying negates conflirge, thus when women do not want to answer, this is always, because, the unfavourable truth will result in being out right, out her bias, which is conflictive to the asker's bias.

Evaluating our attenuations, until we bring out our internal attenuations cause, is called edukatus (education) what we take from education, is what we ourselves come to learn.

We learn about ourselves, via edukatus (education), and we learn for gauging the world accordingly, to what we learnt ourselves, via ourselves.

Mom over here

Intro

We all definitely, (in twenty twenty) have indeed a global conflict bias, which due to the pre-existing conflict bias, none of us, except me, have inner stood.

And upon recognising, what I have come to understand, I have come to realise, the books and videos I create and write, no one else on earth possibly could.

The most sacred bond boys experience with respective mums, could be deciphered, and written about (as I intend to do) in a major body of doctrines, such as psychology and child psychology and sexual psychology.

Main

Explanations of mis-externalised sight exchanged between mum and son, has never actually been seen, as a, otherwise, apart from me, men haven't seen via women's eyes, or visa versa, who haven't even seen via the others eyes means neither knew how to perceive, to in tern conceive, why men and women, world wide miscommunicate, except for me, I have, thus I come to learn, unperceivable and thus unconceivable sight, all women's possess.

And I am teaching the world and everyone from my YouTube channel, the secret women speak with.

Men have never seen via a woman's eyes, nor visa verse, thus no one human has summarised, the results of possessing, both women and men visions, at once, to than be able, to translate for men, and incredibly enough, for women also, what the other is doing, exactly, as of yet. So far no one else has done so by learning to use two sights, in one body, mind and brain, so could not translate to the world women and men before me.

If anyone did, someone would have told, all these women, men aren't looking at them, directly per say. If anyone knew what I do, women would have told their sons, already.

Outro

48

"Mummy, can see you, when she's facing ninety degrees away". No mums told nor shown her son, how to see when women are looking at ninety degree angles. Both visions are needed to translate the application, of gauging ones mum (all women) one's (son) respective significant males.

So no son can get to just listen to their mums simply telling the boy "Son, I did see you ...". Boy's (son's) only get, about what's visible according to their sights, own optic signals to their own bodies, is what their biased definition of sight is. If moms not calibrated in that framework, boys won't recognise mum, as seeing said boy. This isn't possible for said boy to see, from such angels, unless he's been doing traditional Goju-Ryu.

The neurons rely on sight, as the supreme authority and hardest form of evidence perceptible to man and women and child and animal alike (mammals).Sight is used to drive around busy roads, full of school children, while driving, which is learned via sight, sight is used to drive around all the children.

This isn't mars and Venus.... This is just peripheral and linear, vision, and behaviour, and concepts, and goals, formed either peripherally or linearly.

Intro

Use your real eye's to realise the only "Mars-Venus-pattern-of-conflict, is really our eyes-sight's, yin and Yang, and so are misunderstood, from each others points of view.

Thus nothing can give more concreted definition, to our perception of what is really real reality, than sight, thus the boys neuron, tell his mind, via sight, how to define sight itself. And men's neurons don't teach men how a women's sight works, so men have to learn it.

What sight really is, and of course, everyone does this, with their own optical neurons, doing the teaching. (Its not the father, never is with un-socialised boy's, they also compel sight be administered to them the way sight actively actually appears to all boys, yet not to any girls.

Thus grown men fight looking at women at cinemas do not actually believe when women do so see men looking at attractive actresses, just because the women are facing the screen, directly, from ninety degrees away, women see men and all else just perfectly fine.

Main

The optic nerve conditions and programmes the body about what sight is, so men (and men also means all boy's and son's) have to see through women's eye's, to get what women are, only seeing and not showing, works directly on the neuron, to teach the body directly.

Mom's need to see through their sons eyes to see why men are so demandingly compelling head on attention from mum all the time.

Obviously, both visions are needed to accurately translate, the null void of misconceptualisation communicated.

No one has, so I know the best appropriation for my knowledge, is to evaluate how coming to be attenuated to this knowledge, helps support

50

love on earth, for countless human life cycles to come.If I reincarnate, may this knowledge (and gallery,) be available to aid me in my existence, be I born a woman or man.

I don't read dating books, I have seen through women's eyes, I see alot. Thus I grasped otherwise unattainable / un-learn-able insights out women's vision. Ki

Thus allowing all men to know why all women, even men's mums, don't look, directly at them.

I have to show the world, one book at a time and one step and one video and product and website at a time, how to calibrate the opposite genders behaviours. Visually, my gallery is explaining women's language, so men can't afford to miss this, nor can women.

As only the sight, needs to be synergistically, switched around, to gain this conceptualisation, once acquired, it lasts forever, and can be a vital skills to obtain, even supporting one's survival.

Summary

No one else world wide is going to know this about men and women which make all humans, glancing or about the mis-externalised visual perception between son's and mums. This first time this occurs between men and women, and it goes on, forever, unless we learn now.

Nor can any other authors tell men about the astral unborn baby, women hide from their potential suitors, at parties, in a mini skirt.

Nor as I am about to, can any other author, explain, that women, love only once in women's entire lives.

Nor after women's virginity taker, that the only love a woman can ever again possess and have to give away, is love, for and from their child.

Yes women fall in love with their first, and never actually ever, come to love again, in the same incarnation, except for the love women have for their own children.

Outro

This is unknown to all dating experts. Unknown to all psychologists. Unknown to all, except for being known to all other women themselves.

No man world wide, can give that to you, information I have come by, this information is fact, and should become pragmatically integral, by way of being useful facts, no body else, will generate such information.
Were else can you say you learned women have one love per carnation, in a male partner, no one's ever attained and taught anyone this, about women's virginity and women (in)ability, to love any men there after is no existential.

Yet once obtained, each person should be able to work out any women at all, with ease and visa versa, any women to decide any men's behaviours with amazing ease.

Like when women have been cheated on, the partner, seems black and dark like an alien entity, the room oscillates, like a vibratory disruption occur, in men which cheat, when men's women look at them, within seven to nine days, after.

As soon as women see men, which have cheated with in the week, women's normally excellent sight, goes awry, thus women say to themselves right at this moment, I must be crazy, and women thus convince themselves they are crazy.

Women depend on sight for survival, thus women's sight going awry, convinces women they are going crazy, thus they are basically, not seeing, their partner, is being perceived correctly, as women have the best and better availability of sight.

Women may now interpret this spectral visual awareness, and know a man has undoubtedly cheated, if women go with a bias, to how women are used to seeing men, which haven't just cheated, women will successfully convince themselves, they were not cheated on, and yet women will innately always know, convincing themselves their crazy is the, rational response to knowing one's been cheated on by another without validation, yet not the accurate response.

How much does the sense of in sight define all behaviour, all speech and the meaning of words themselves

Intro

Yin and Yang

Why not if this mis-externalised, all comes down to sight-gained-biases, causing conflict amount men and women. Notice women are alot better at not p***ing other women off than men are, this comes down to awareness, women have loads and men are not able to exchange with women were women's elevated level of awareness.

Main

Smell doesn't produce, many conflicts, nor do tastes, nor touching, as a sense, yet sight, as a sense, produces virtually, every conflict conceivable.

If the streams of sight are juxtaposed than the behaviours formulate conflictively, as is the nature of being synergistically-juxtaposed. Juxtaposed sights, produce juxtaposed (perspectives and more accurately, biases)in France etymologically).   Conflictive perspective's Induce conflict, people protective perspectives over their beliefs, leaves people unprotected themselves, to protect their perspectives.

Outro

Virtually every bias, which is the grounds, for conflict, is born from, the sense of sight, thus sight has, created the behaviour men and women each have, and the interpretation escapes each respective gender due solely to each one's own, pre-existing, standard of sight.

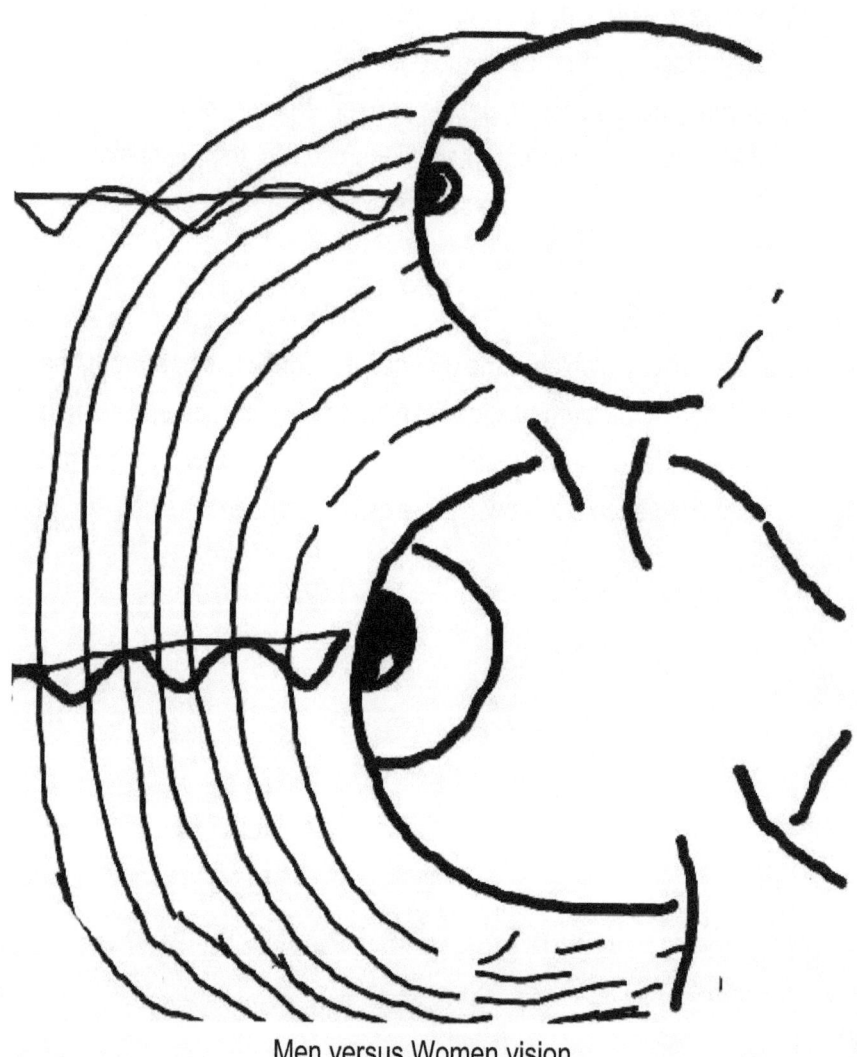

Men versus Women vision.

While the men (top) need to face things and people to see them...

Women, (on the bottom picture) needn't face things and people directly to see them, thus women find men, quite offensive when men, let everyone is society know by staring at them...

Between women and men, conflict biases exist, just clearing this up for all men me women

Conflict and biases-effects

Intro

Examples of an conflict bias: which demonstrate: the collective cohesive power, all humans, this variable, is one universal timeless example, which overwhelmingly evident anywhere in world wide equation, which is, women interpret all men (regardless of culture or position) which glands for just a momentary millisecond, as sexually signalling them, this attenuation to the conflict bias, women's juxtaposed sight, lends to their conception, of men's sight, causes women, world wide, to utterly mix read , any glance a man directs for just a millisecond.

Main

Women world wide sincerely believe, every women does, that these glances are definitely sure tell signals from any men, which their received by, one millisecond is all it takes and any women any were on earth, thinks your sexually, into her, based on a split second (probable) glance.

Even when men are looking past a woman for a split second, and not stopping to actually take a look at her, women, see men, as looking sexually anyway.

Even if the woman's, five hundred years old, men glancing for a split second, is still universally, amongst all women, seen as a sexual symbol of interest.

This occurs because women gauge their environment via juxtaposed vision to men, this sight, gifts women with totally complete spatial (and visual) awareness. so women don't have to swivel their heads, to see anyone by directly facing anyone. Men thus can not correlate, women's behaviour visually.

Visually being the same men correlate men behaviour just fine, women are visually able to internalise what women see other women doing, perfectly almost.

Outro

Men, cannot get any casual (tight) correlation, when men externalise, men's own internal way of seeing, or when men externalise their behaviour and actions, fashioned out of their sight.

Or globally, boys would not demand one's mum's to look head on at them at them, and what we would instead see world wide is that all boys wouldn't conceptualise of being looked at like that.

If women could tightly correlate women's own innate way of seeing.

Than globally, the way women talk, all girls would know why boys (seemingly glance), and sons would be told by informed mums, I don't have your sight.

Yet globally, it is evident, that our standards of sight, don't allow men to understand women behaviours, nor vice verse, women externalisations, from the same reasonable presumptions, projected from each other's first hand experience.

This is firstly, a order of perceiving and doing things, based on the sight fashion respective to gender.

Men, have all seen
this face.... On women at least, when women Look at men, men do not do
this, after a split millisecond of a glance (0.4 secs), because women see
men first, and men have to look to see.

When women get a millisecond of a glance from men, women do this
OMG face, why do women make the "oh my God face", women are
already presently looking at the man which glances, which gives women
the chance to get validated, by all other women around (remember
women are narcissistic), due to women's vision and conceptualised reality
and women's behaviour, which comes out their female vision.

Women thus get recognised, properly, by all girls around, men's vision,
looks around thinking he's in trouble, yet women, glance to communicate

61

sex, so this face, means definite prof, all women, want sex, from a split second glance.

That's what they are externalising, it fits the bills perfectly, doesn't it.

Men glancing at women, collective (sexist) gender bias.

Not pushing an agenda, these words just felt, upon evaluation, the exactly correct words, does anyone have a dictionary...?

That's able to see each other yet... What about now... can you see me yet, can you see me now.... Now do you see?

Intro

Men do not get why, women believe this, because men have to turn their heads to see anything, and don't have sufficient sight to see an environment all at once.

Main

Women mostly believe men see said women, men mostly believe women aren't looking Men over, their absolutely oblivious mostly.

If the two can not correlate when each one is connected visually.

If men and women don't know cues and signals nor when their even seen, or visible, or when one's behaviour, is actually anything, but inconsiderately fashioned.

The bases of juxtaposed calibration respective to the opposite sex, (of, well, the entirety of) all the world, means undoubtedly, this will be the basis, of biases to come against men of all of the entire world over and over again.

The same biases are based in the same rudimentary qualities of sight, thus producing the same biased based on the same modus operandi, until sufficiently calibrating the opposite sex, this leaves approximately 3.5 billion out of approximately 7 billion lives, which one may be able, to accurately calibrate or rather cross calibrate, without misexternalising, while knowing what a woman's language means.

Even were unpleasant, truth is essential, men do not want to go on without knowing the essentials, like if you sleep with her, she will not come to love you truly, only her first (virginity taker) is loved, every other man it's conditional love, for children, unconditional love exists. The man may fall in love though, so is at a disadvantage.

As women first commitment to have the baby with her first lover, goes out of the window, women always love the man, yet the focalisation, doesn't reoccur, until pregnancy and conception, every other man is not loved truly, this means men become the disposal member of any relationship almost.

Outro

The only way we become aware of bias, is via conflict, by which conflict gauge's biases, once differences become to great, fights may occur

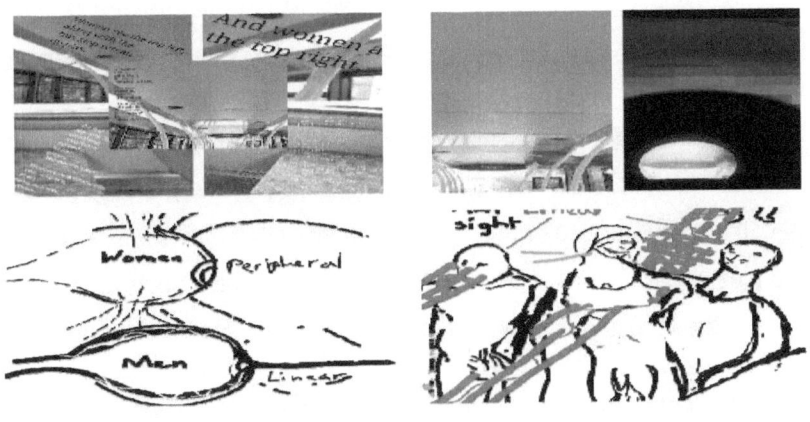

"He's inconsiderate", so she tells me...

Men sight means men's behaviour is not able to encompass as much environmentalistic assessments and in tern not as many considerately thought out actions nor structured sentences, according to any given environment, bus; train; park; pub: party, women have the vision for 90 degrees each side, a full one eighty, for an unwavering reference point, than a little more.

So women are considerably more environmentally aware than men, so get readily angry at men, for seemingly being less considerate, of the surroundings.

Men obviously can not compete with women environmental awareness.

Women obviously cross externalise, women sparse sight, on men, this presume, this is something, which men could actually change, and become aware, environmentally, this comes from women with up to forty fold plus, more sight than men, when men start to think about a thing they are focusing on, see the image of the bus display screen.

Women getting angry about this, are no different, in any way, from little boys, getting angry, about mum not looking directly, only, when women re grown, and doing this, this is why women end up on death beds as old women, with no men.

Because the sexist prejudices, of presuming men would purposely, not do things, out of just not giving a damn.

A woman, the gallery explains everything men and women need to know, to love harmoniously, if they can be free adorable at all, not every man and woman can be.

Basically, men are not visually, able to see enough, see the collages of images above, on how much more aware women are than men.

,

Men aren't as visually aware, see left collage of images.

Females are sexually discriminating against men, via projection, of a narcisstic nature, just like the gender based conflict biased of men, glancing ,being taken by females, all over the world , as a sex signal, taking their sons, for narcisstic, for asking their mums to look at them, women than accuse partners of cheating, based on women having the above sight

Women than get upset about men glancing, thinking he's disrespecting her, even while she has look everywhere.

When women say pay attention, women are sexually discriminating against men's inability to put the environmental awareness to the level of a woman's, which is virtually, impossible.

Women need to be told, look, I have men eyes, look at the picture, women won't give. Damn by the way, women are narcs, they want to abuse men, and not provide any remedy, this way women have a real dog, and they get to abuse the dog, because women do not actually love men.

Women do give a damn, what women's own eye balls teach them though, women will believe, when to use their eyeballs, women have to look, via head turns, and facing people directly to see, women will thus realise, how men have been looking and seeing, and see women are sexist abusers based on their sight being vaster, and sparse, and ultimately widely peripheral.

Let's see said women, with a balaclava hood, even tell you to pay attention in public, let alone, do any better than you do, she's externalising her vision as a projection, this narcisstic trait's, the reason, their own vision, has to correct, their own neurons, processes, to calibrate, correctly.

Sexist, means power to impose, on a discriminatory ground, meaning women are being sexist.

Women are also misexternalising/projecting/cross calibrating, yet men have no concept of the world around them, the above images show women, what men see on the left collage.

With this balaclava hood, she will not look at every man's crotches and will, thus, know men are not looking and thus, chill out on men.

Men simply put, your not meant to be even almost s environmentally aware as women, in our modern state.

The above left Vs above right image, clearly shows why and how this is women's sexist abuse.

68

And stop accusing them of infidelity, and she should watch movies with telescopes at the cinema with you. Because than she is really going to enjoy the experience more.

If men have never seen, there's never been, any reason to structure words considerately around the unperceivable to men, which have never visually seen a thing. Meaning, inconceivable, as men can't apply nor strategize, with anything at all, Men can't neither sense; nor identify, women perceive everything, a man fails to fulfil to satiate considerate action.

.

Women do identity the things, which men do not fulfil, which women feel obligated to comply by, this means women's sight, makes women, more bothered about unconceived of aspects of a men's behaviour.

Inconsiderate?

Basic courtesy to other, happens at almost 360° all the day long for all women on earth, thus men having telescopic sight, get engrossed in anything teens sight is on, and men thus seem uncaring, cold, inconsiderate and unaware to women.

Peripheral vision
(ALL) women

Wo

men vision versus man's sight, she's superior, if she were a survival team
member she observe above her entire male squad, and be more aware,
in an zombie survival situation, I would thrive from such excellent intel

Don't take women to the cinema need on the above image, the movie is not a focal experience, like for a man

Intro

Men are guilty of conflict biases also, from boy's to men

Equally, boys make their mums, focus head on, were as women naturally see perfectly well from ninety degree angels. Women have a problem with conforming to look and face head on, this is the only angle which is used for conflict in females body language, and also, the only one men seek to occupy.   Sometimes women mistake foreword facing as over engagement or aggressive, unless, she's developed oxytocin behavioural traits, such a faithful comfort ability.

Main

Now when your at the cinemas, maybe, don't look at other women on screen, because, even facing forward to the screen, girls can still see you, perfectly clear at the exact same instance, without turning to face you, they interpret your look as sexual signalling behaviour to the attractive actresses on screen.

Once a psychopathy, of psychosis, has occured, they spread by participating in tradition.

The conflict bias, perpetuates on a global scale, and on going basis: often humans die before being able to decipher the existence of this fundamental schematic: and so this goes untaught: yet one can attenuate virtually the whole world around to support one in a non intrusive way.

One may effectively, effortlessly, check if those around them, are in support of them and their actions and behaviours, and more importantly perhaps, attenuations.

Other's deed's acts and behaviours may or may not be in support of one's goals, one can effortlessly, effectively and easily, palpate for this such conflict bias in those around them.

By presenting an aspect of the representation of one's attenuation, and enjoying, or engaging in this before them, if this arouses conflict, their bias, is directly adverse to the fruits of your attenuation, thus averse to your success, be this directly convey or passively aggressively displayed (insidiously and gradually).

Biases surface, because the neurons are disrupted and digestion shuts down, and vicarious fight and flight responses occur, passively and directly. Yet activity, and restlessness or ambivalent trajectories to one's own indicate this.

Don't split with a partner, before you at least confer which the syllabus which converts, your interpretations of your partners behaviour, with what's actually occurring, respective to your partners gender.

The sight boys and their mums have are as different from each as tin is different from Yang.

Yet all the lives on this planet earth, presently equals seven billion and previously, billions and billions more have lived, yet, all mums and sons, born world wide, have not worked out the reason sons accuse mums of not looking and compel mums to stare, is because women see totally ambivalently to the way men see.

Summary

This evidence facts, reflects itself, when women world wide, still believe, boys and men, have the same sight as them, thus glances seem to be sexual signals to all women, because, because mom's and sons, never work this out via through evaluation, thus never come to realise, they see totally differently.

And their in lies a bias, which has been occurring since the dinosaurs, and is still occurring, despite the presence of the internet to explain this overwhelmingly evidently global phenomena.

Shouldn't men on dating sites, see fit to ascribed this kind of insight, so one may start evaluating on a basis, which best allows self growth and development respective to reproduction, an essential social economic aspect of survival.

Outro

Yet no one does, no body will, so I have to put forth the absolute limits of men and women, otherwise you, yes you, won't, upgrade, upload and update, the planet.

Men are sending sex signals all day, and women are sending men sex signals all day long, yet men, have no idea and women have a collective psychosis based on falsehoods.

Women are fighting other women, in physical violence, based on believing and becoming jealous, about a (psychosis about) perceived significance of a males glance.

Women may be, somewhat abusively, insightful of conflicts between men, women may want to see men kill each other to choose a victor, and getting their partners to compete, with men they perceive more dominant, so women, may choose the victor, and the men, often have no idea.

Except for an inkling or a sixth sense, men, which may go through a lifetime without having a single fight, may make men empirically, maybe notice, that while around, some women, the certainly unlikely likelihood, of having dangerous fights, (and conflicts), go up significantly.

Or only occur within certain women's company.

Based on sight, which one finds more reasons to attack a partner, which is a Marc, which is a cheater and a liar, thus negates culpability for attacks, which one stands outside the pub, screaming at fifteen men, single headedly for an hour, without getting in trouble, what's that woman going to do when she gets home to her man, and him up and maybe triangulate power at him after, even contrary to popular belief, what's the overwhelming evidence pointing to.

Intro

Two partners, one beat up the other and called the police after to get them in trouble and succeeded.

With out ascribing gender, we all know, which one got which one in trouble, despite being the only one guilty of violence.

Act like you know.

So contrary to popular belief, let's state what's evidently the case. Despite what people believe, coming to evaluate (from mathematics, meaning taking the value/s;- of) our saintly beliefs around women, we see women are chiefly manipulators and skilled liars and actors.

Ne-gat-ivism is a natural female way.

Main

Women may get men to fight one another to choose victors, and yet play damsel in distress, all men will support any women seeming vulnerable and distress, women fake this reaction to get support from everyone, often to avoid having to tell the truth women are lying about, or not being straight out with, as men will fight for women and yet all will fight to protect women, thus this is ne-gati-on of hard information for evaluation and a heap of arbitrary, enigmas, which equate to an upset female, females only communicating upset and discord, while usurping real actual male protection and accessing power there of and always lying and manipulating men there of to do their dirty work.

Badgering or checking with women for real hard facts one needs to evaluate on, results in women ne-gati-ng to speak up and answer questions, (which means what?) Women negating to answer questions don't have favourable; honest; truthful; words: unlike speech wrapped in enigmas allowing women to usurp men's primal urge to protect nurturing women, firstly our mums.

This makes men bias, as their protector instinct is conditioned to tell them what to be biased against, and so of course not evaluating, and being in the same collective psychosis of all men globally, which see women as vulnerable, which contrary to popular belief is not so in anyway, except for one way, which is of course men's physical dominancy (over most women, definitely not all women). Women will just move on to other men with more dominancy, power (, fuller fridge's) and loyalty to their women.

Outro

Women are never in danger, scarcely ever does this occur, men however are frequently endangered, via women getting men to compete and fight or kill one another via conflirge.

Historically, take the historical knights on horseback as an example of women collectively, getting men to slay one another or presently injure or mame one another, as a evening of pleasure.

Knights are already Nobel's with status, perfectly good knights with horses (a symbol of wealth), when women already have a dream man the vision women have, causes the female eye to still filter through countless other partner's, while men can not look at even one other woman, without knowing, a jousting tournament is just around the corner for them.

Presently, a historically similar relationship between women and men still exists today.

Women take a reading of men's faces to make certain men are still loyal despite the conflirge and visually validate men, women evidence an awareness of danger and scarcely will women get their own hands dirty.

Or calling police on fraudulent ground to just attain more power, or jousting perfectly shiny good knights to dismay and death. Is why a damsel even gets in the distress position, women can solve their problems, instead this is a primal trigger, which conditions men to take care of women.

While being protectively cared for, women will get men to compete and even kill or get killed, whilst in his protection, the man's kind of like a dog, she's validated for getting to fight for her.

Women will get men killed just to get real validation. And props from women and not so real validation, from actual men, yet women's mis-externalised global perception, means one more MIS calibration, about what (virtually) all men actually think and feel, beyond the thinking's condition.

Men tell the truth and are clueless, as to how women's non-focal (focused) vision, teaches women.

A woman's vision, makes women able to be so cold.

A little small amount of alcohol, allows women to lose all sense of men's sensitively focused mind and sight and men's often highly (focalised and intern fixed and so) delicate emotions.

Women don't care really, if it's not them going through it, this is to say, men are easily more compassionate and caring and sensitive to all other's.

This is why although all men will take up and fight for women, at a sign of distress, women will not, despite being significantly more aware, will not, women won't come around to fill up your empty fridges, where as men will.

Problematic patterns can be corroborated as it happens

Intro

Looks like men and women have loads of consistent different visual and intern mental concepts....

Consistent defences, which differ, congruently to men and women's visual differences and intern their mental (psycho-social-spiritual and also emotional) make ups differing.

Women may be projecting collective visual biases which women worldwide all also do.

Equally sons push on their mum's and in tern all other women's, sight standard to fall into men reference for calibrating sight.

Men and women world wide are globally pushed into envelopes of behaviour and sight which neither, women nor men naturally nor originally belong to.

Rather women and men respectively, belong to their own envelopes of visual (and so visible) behaviour patterns, thus men and women respectively, have separate patterns, which are self forming, by way (in accordance with) men and women's respective levels of sight.

This means, being able to see one's own world, is the way how one comes to externalise (forcefully even, from a stand point of a bias), that the entire world, see like them.

Now one see the world around oneself, the only way one knows how, one '(s)own standing, thus one 'sown perspectives.

Main

No one, at all, crosses the road, from another's perspective's, physiologically nor psychologically, also, as psychology is only perceptible, via the physical, no studies nor experiments have ever been conducted on the psychological, which weren't directly based on the physiological.

Meaning everything one's, physical experiences, allows the psychological to develops out of one's experiences.

This gathering of experiences, leads one's physiology, to gauge, one's surrounding, with one's gained experience.

Yet never using the sensory capabilities of women versus men or visa versa, means only one ('s)own perspective, is utilised to come gauge one's world, around ones self.

Meaning despite vague mention of women having more peripheral sight than men have available, is just a vicarious, lifeless fact, with no bearing on men and women.

Because, vicarious information, is very different to everyone world wide, versus non vicariously experienced sensory information, such as the sensory information, which each human on earth utilises to perceive the world around oneself.

Meaning, even those conducting these studies, have clearly, themselves miscalculated men and women respectively, in their own environment, as even researchers, are going based on their own standing, so respectively, men seeing juxtaposed to women to the extent neither makes communication aptly, means, men and women researchers in their field, do not know why sons and mums, mis externalise.

See until one truly sees via another's eye's or sense via another's senses, one does not truly obtain meaningful information,.

thus can not truly educate themselves, why three point five billion men (males) can't get (women) their mums to look directly , yet every son, struggles with his mum about this exact issue. So thus men and women researchers all have this knowledge of peripheral visual, women possess, yet none of them, not one, sensicalises, the bearing which, the physiological reference point of sight, converts, conditions men and women for respectively.   No psychology doctrine of practice could educate and calibrate psychological phenomenon which are highly consistently occurrent, without explaining, via each respective woman and man's perspective, and explaining their truer calibrations, to allow women and men to gauge one another other, because so far, they have significantly miscallibrated interpretations, of one another, all around the entire globe, right now, based on overwhelmingly global evidenced by this book, so far, you can see this facts, upon evaluation.

The vast majority of men, all women have ever seen, have been wrongly misinterpreted.

Every son born to a woman, has greatly misinterpreted the visual information each son had seen and processed with his projected vision, on his mum's behaviours, world wide, today.

The only visual interpersonally communicable interpretation, men and women world wide derive from one another, respectively, is wrong.

Men and women, from current and past incarnations, definitely, misexternalised, women. and men, mis-depict and misgauge, what (and when) women and men (actually) see.

The conundrum of all of this is the only way men and women can see the way each sees, is communication, and exchanging information, which seems, pointless to explore, because, a bias of sight exists, and as sight gauges our surroundings, we assume, sight automatically, produces fact.

Empirically, evidently, men and women, each autotelically and automatically believe, via sight gauging ones environment, one also has gauged all others people's sight's, in relation to our own standings.

Outro

from gauging through each man's and each women's own stand-point-of-view.

Put another way, virtually none of us men nor women, will start a car with our eyes shut, and continue through out several journey's, without opening our eye's.

Thus we all (virtually) each of all women and men, take our own sight, to be the hardest of rawest evidence, conceivable.

Men and women imagine, nor conceive, of any evidence harder than their very own sight, remember this important fact.

In way of a bias, women use women vision and men use men sight, which lends to neither ever crossing to the side (bias) of the other.

Thus no psychological expert, every knew of this phenomena, as neither men nor women, conceive, of anything beyond visibly conceivable representations, such as having, the greatest bearing, on their present existence.

Thus the most obvious apparent factor to anyone which just absorbed what is written above, knows, this kind of discovery, to unravel women's entire behaviour sequence, precisely calibrating present live behaviour in the here and now.

Do damsel's in distress only cry wolf to shepherd the sheep (men) into beefing on women's behalf.

So women can be abusive, and even violent, while negating getting their womanly hand dirty.

And if men were the aggressive ones, would men not redirect the strategies attenuation, passively aggressively, or aggressively in a direct manor.

If the women destroy men and get other men destroying one another, for the pleasure to choose their hands dirty directly, men would notice, yet with higher dexterity, women are imperceptible, thus I had to create a YouTube channel to just elucidate, how women are perpetually communicating all around the world, this is the only way any woman communicates, as there secret language women speak with is visual originally for every woman born and becomes a secret to every man born, due to his own sight.

Sense of sight

Sight is our(teacher) sensei sense

Intro

Name one thing you ever learned with out sight, and if you can not name one single thing you learnt with out your sight specifically being the way you came to learn.

Sights responsible for the entire world all around any one of us, and the mental concepts drawn from sight are equally, the world around us all.

Vision may decide entire careers, whose selected to make a family with, who's the world champion, and which nation we live on are all coordinated via sight.

Smell taste and perceptual touch, all combined, do not have the opportunity to fashion our surrounding world, a fraction as much, as sight does.

What we see is considered the realest evidence form perceptible, anyone possesses.

Main

You are exactly the same as virtually everyone on the world , we all use
our sight to learn, and even to gauge (perceived) facts.

The hardest strongest evidence to all men and women is their own sight,
seeing something, means men and women, know something is real.

Men and women thus in tern, drive with essentially their sight, and avoid
killing everyone around their vehicle, via, the hardest form of real
connection with one's surrounding, which is undoubtedly, sight.

Thus all men and women, takes their versions of sight, for the hardest fact
to evidence, what one's own sight actually really is, and thus to conceive
sight, is the opposite to conceive the opposite sexes sight, which no one
else in the world has done, ever before.

Thus I rationalised, obviously, millennia and aeon's, would elapse at least, if not much longer, before another one of all the world's women or men, elucidates the global collective psychosis women and men respectively belong to.

Now we know we all come to learn our house number, address, or date of birth with sight (by way of sight). Thus we all come to take our sight, as evidence as what sight actually is.

So when someone comes along and says, "women have a bit more sight", that will never make any one, see significance, from one's-own standing, look presently, around the world today, this vicarious fact, has not.

Until you explain something from one's personal existences automatic interpretations, without visibly absorbing for and via thine own eyes,

Is the difference between telling some and getting one to do something them selves to internalise the benefit and teachings.

If I explain a karate class over and over for seven years, you have learned about an unusable conceptualisation for those spent seven years, which you never get back.

Thus getting you to do a karate class for three hours a day, will allow you to internalise and develop and go beyond limitations, with consistency, even defend yourself and also compete and extend compassion, comradery on to all others.

Outro

No one will, actually, see, as one sees and in tern can respond, from someone another's, stand point of view.

So be sure to visit, the secret women speak with on YouTube and my website today to start learning right now how women all see, this can be used to decode history and even sensicalise murder cases.

What a woman sees through her eyes.......

when she's sitting between two guys,......

for the record, virtually only a small amount a guys know this, other than me one other guy, born the same day and year as me, virtually all other guys do not at all know this, uncommon fact, yet popular truth.

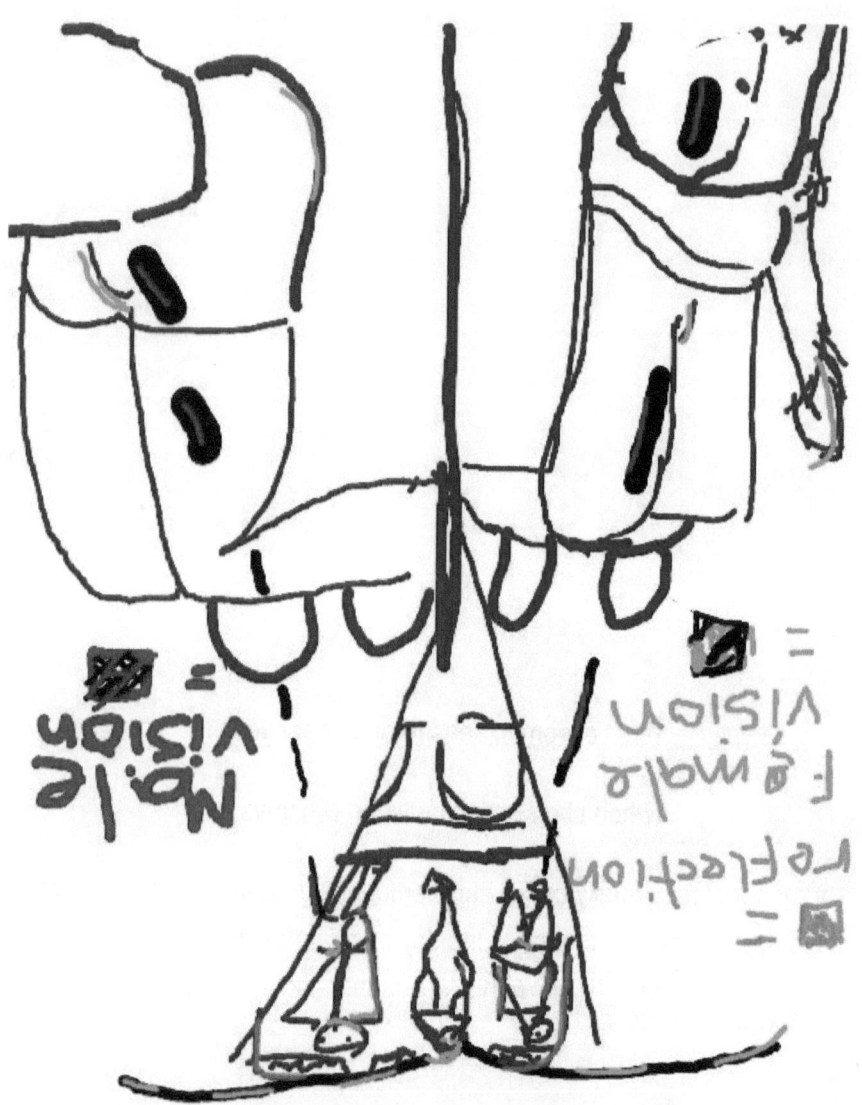

What a lady can see, (non-stop) while she's riding that train with you.

Range of vision, pink and black lines = peripheral

Blue lines = linear male sight

Introduction

I have come across accessing women's sight, via practices, whichever would probably kill people, if they were to start trying to do what I did to come across women's sight.

Empirically we all, we had no idea, before these writing and my YouTube channel of the pre-existing juxtaposition of men and women's sightseeing, leading to almost all other differences between women and men.

Empirically we all know we believe we have, we haven't heard anyone in the world ever, say to anyone of us in the world, your son has a tiny ability to see and thus, your son's, seeing what believes sight (itself to be) and is getting his mum to use (what he is positive) is her sight, to be able to see him.

What is sight really?

As the conceptualisation of what sight is, is drawn for all men and women, using our very own individual sight, respective of women and men.

(Son's and mum's:) Men and women, see differently, due to the factual realest, hard information available to men and women, the visual kind of information, which is sensory.

Actual real reasons, which is men and women (see whatever sight they have been using to see with, as what sight is itself, is due to men seeing, via a member of the opposite sexes eye-balls, directly.

Seeing through one's eye's, activates one's entire brain, tailored, to what one is seeing, the neural wire, which fires, conditions the whole body, that's what visible input (known as sight) does.

If sight sees a hot women, a gun directed at oneself or a tiger approach, or a car or a bus, one whole being is being conditioned via sight, as the neurons fill up every aspect of one's whole being, with the visual information attainable via seeing.

The neuron also forms the bias, respective to men and women, based solely from sight itself. Thus men believe, sight is this shafted, linear visual focus.

Women will believe sight (vision) is a awareness sense, for seeing misty degrees, to either side at once, with pitch perfect focus, yet not of one point of context, yet to all visible aspects of the environment and of one's own self.

Means the bias the brain of all humans, has, about what sight really is, doesn't program a schematic for women's sight, as the neurons, the optical neurons, for sight a

Thus the world population of women thus believe, all men, want sex, when so ever even a millisecond of sight is glazed over a woman, which happens to be in one's surroundings.

Now we can all why this is the impression women all get. Its based on all women forming their only factual real concept about what sight is, via using (their very own and individual) sight itself.

Resulting in women, which hear the word sight and vision , to lead them to equate the word sight to mean, only what one them self actually see's, as what sight itself is.

So men believe women are not looking, unless women, direct and fix their head on gaze, at men, or seem to , women can move their eyes away, often as needed, yet fix their bodies facing directly.

Than a man will believe you are looking and thus seek to extend the visual cue, by getting to know woman, directing her body and especially eyes at him.

After all world wide, what else have all boys and en equally, asked for, of their mum's.

Give all men, exactly what all men have already all asked for, and if you do, your chances go up, through the roof, if you get good, your response rate from men may go up fifty fold.

As men get one fiftieth the directly focused attention from women, based on seeing, men all conceive what seeing itself entails, from the neurons, which themselves, make men able to see, with these neurons, the bias, forms in each boy and man's sight about what sight actually is really.

Women world wide, seeing with vision, developed a bias, derided, from the same optical neurons, which produce vision for the women's central nervous system, thus these neurons form a biased image for women, about what men are doing, when men use their own sight and when women use their own sight, men never conceive, what makes problems arise, out of things men can't track nor see (so don't conceive).

Really seeing, with one's real eyes, will allow one to realise, women /men real observations, in the present moment to moment basis, and calibrate according, to what all women and men are really conceptualising.

new mates from out of the conflirge, upon evaluation, which is more caring and which is more aggressively orientated to violence. Which is skilled most at manipulation, and thus is optimally aware, and thus which, men or women, which one out of women and men, have to be able to innately, based on sight, be more negative (dishonest)/disallowing, their actions, to get directly, back to them.

Men couldn't trick women into fighting world wide innately, as a common place behaviour trait, as women are too perceptive.

Women are the only ones which get men to kill or almost kill each other, to say, here, I will pick this one, (if the male is so strong) even though she chooses a victor, they could kill her for mutant and treachery and basically, having no bearing of the nature a relationship, should be based on, thus a woman which impregnates with a men she indoctrinates such ways, will obviously, never be loved by women world wide, only the virginity taker and the women's children and then women themselves, will get any protection from such pointless, fatally dangerous conflictive situations.

If women went around getting

Have knights in shining armour, in love with damsel's in distress, ever encouraged their damsel's to joust to death, slay strange creatures, spill their hearts out in cages, or fight every other attractive woman, historically, knights in shining armour have not, yet the conflict bias of all men, based on "protector instincts" demonstrates men as the more loving of the two, women love on only these occasions, and virtually, none other than these ever.

Women love virginity taker, forever and always.

Women love the first child always and ever more

Women will always love their children

Women may (not necessarily) love their parents (mum and dad)

Women will not ever love any other men besides the men women lose their virginity to

This leaves one to essentially one's own perspectives and one's own psychology as a result, psychosis, is the misrepresentation (mis-externalised) calibrations for one's relevant reality.

and the only way, which one comes to start gauging, one's own individual key word being one's own 'individual', world.   Which is what we are all doing.

Now with seven billion people doing this, and at least three and half billion of those people, have always seen a totally different world from all of us.

Than no wonder, the world, and all the people on the world, for their infinite potential, light to learn of the existences of their conflict biases, so if, one wishes to explore via evaluation, the state of the whole wide world, or just something they know they know they have not evaluated and instead fixated at conflict biases, one may decide based upon evaluation to explore expansionary attitudes and goals and even track certain feelings and sense included at certain locations or around certain groups of peoples.

Men do so externalise first on their mum's, women equally also externalise the world by what women have themselves come to gauge their own world's with.

Go on YouTube, and stop reading books, learn quicker from a lecture or a image let alone a video.

My videos are one a kind, kindly like and support my videos.

YouTube channel GOJUVISION

YouTube channel SECRETWHICHWOMENSPEAKWITH

YouTube channel SECRETWHICHWOMENSEEKWITH

YouTube channel secret women seek with (visual video and evidence of this real language women speak, so men can see and learn women's language, never be received or anything ever again, some women are after fortunes).

YouTube channel secret women speak with the secret decoding of words and signals, non verbal.

Video evidence of the set ups, which women will consistently behave the same in.

Between sight and speech; secrectswhich women speak

Secret which women seek with respectively.

Psychological spiritual astrological social political of women unknown to many and most yet essentially, worth way more than a psychology degree in terms of self education.

Astel
unborn
child

Thank you for reading, I love you all, all of each and every one of you, that especially includes you, your everything. Do exactly what your doing, it's great, your perfect, don't worry.

We are all one, truly, I love you all.

Womans
range of

Sight
via
balaclava-hood

MAN WOMAN